Understanding House Construction

Home Builder Press of the
National Association of Home Builders

Understanding House Construction
ISBN 0-86718-335-7

For further information, please contact:
Home Builder Press
National Association of Home Builders
15th and M Streets, N.W.
Washington, D.C. 20005
(800)368-5242

Library of Congress Cataloging in Publication Data

Understanding house construction / National Association of Home Builders
p. cm.
Bibliography: p.
ISBN 0-86718-335-7
1. House construction. I. National Association of Home Builders (U.S.)
TH4811.U54 1989 89-12476
690'.837—dc20 CIP

Publisher: Home Builder Press of the National Association of Home Builders
Editor: Susan D. Bradford
Art Director: David Rhodes

7/89 SCOTT/PORT CITY 3K
6/90 REPRINT 3.5K

CONTENTS

PREFACE

In 1942 a visionary group of home builders came together to create a professional organization that would unite America's housing industry: the National Association of Home Builders (NAHB). Today NAHB represents more than 155,000 builders and related industry professionals nationwide. Chief among NAHB's goals is the availability of safe, decent, and affordable housing for all consumers. NAHB also is committed to educating the public about the housing industry.

For years NAHB has received requests from the public for a simple, straightforward non-builder's guide to house construction. **Understanding House Construction** is our answer to those requests. It is designed for use as a handy reference, to introduce the reader to methods, materials, and terminology used daily in the home building industry.

The National Association of Home Builders would like to thank the following builders for their technical expertise and guidance in the preparation of **Understanding House Construction**.

K. Michael Cravens
President, Mike Cravens Builder, Inc.
Lexington, Kentucky

Harry C. Crowell
Chairman of the Board, C/L Builders-Developers, Inc.
Upland, California

Ed Dunnavant
President, Dunnavant and Associates
Richmond, Virginia

Roy Dye
Chairman, Bel-Aire Homes, Inc.
Altamonte Springs, Florida

Herschel A. Redding
Vice President, D.J. Redding Company
Rural Hall, North Carolina

INTRODUCTION

Many of us take houses for granted: four walls, a roof, and a door through which to leave in the morning and come home at night. But if you are involved in buying or selling a new home, you will need to know what goes on "behind the scenes" when a house is built.

House construction involves a complex network of building trades and personnel. Understanding the home building process improves communication among builders, designers, city and county officials, real estate personnel, lenders, home sellers and buyers.

Understanding House Construction is your guide to the basics of how houses are built, from groundbreaking to final inspection. Photographs and diagrams illustrate the house construction process. And the glossary at the back of the book defines technical and trade terms commonly used throughout the industry.

CONSTRUCTION REGULATION

A comprehensive system of building codes, permits, and inspections protects the health and safety of the people who build and occupy houses. This system is administered by county or local government planning, building, and health officials.

Building codes

Construction in most areas of the country is regulated at the local level by building codes. These codes govern building; plumbing, heating and air conditioning; electrical systems; and fire safety. While a few municipalities (mostly major cities) write their own codes, most state, county, and local jurisdictions adopt model codes prepared by four major model code service organizations:

- Building Officials and Code Administrators International (BOCA)
- International Conference of Building Officials (ICBO)
- Southern Building Code Congress International (SBCCI)
- Council of American Building Officials (CABO), which is a federation of the three preceding groups

Each group's codes influence a different region of the country: BOCA in the northeast and northern midwest; ICBO in the midwest and west; and SBCCI in the southeast and most of Texas [Figure 1]. The codes are developed by building officials and others with first-hand knowledge of construction practices.

Health codes, which are established and maintained at the county or municipal level in most parts of the country, govern wells and septic systems. (Public water and sewer are usually controlled by county or municipal building or engineering departments.) All builders follow the National Electrical Code (NEC)®, established and administered by the National Fire Protection Association, for electrical work.

Code compliance is ensured through the local building and health department inspection process. Note that inspections do not evaluate the quality of construction—only compliance with applicable codes.

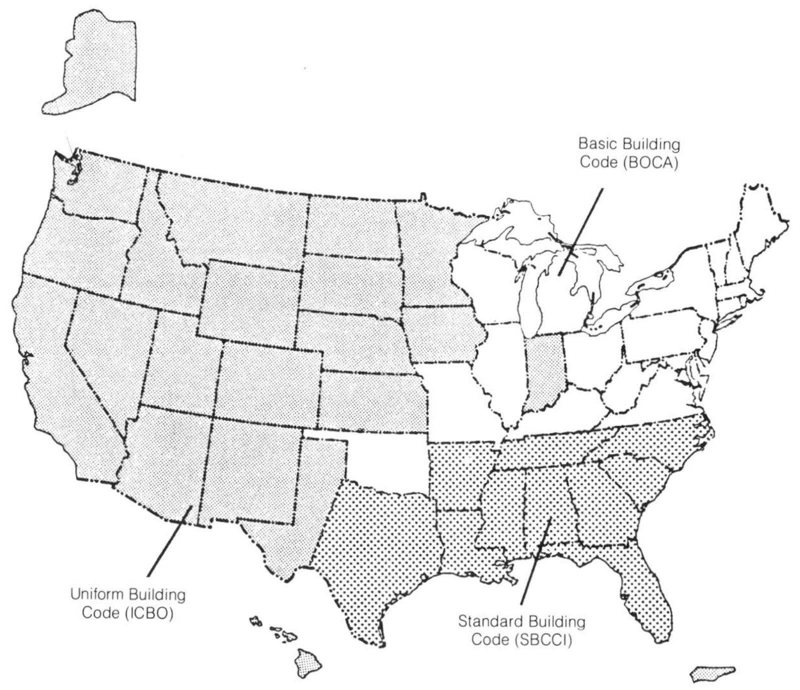

1. Regions of model code influence.

Permits

Builders are responsible for obtaining permits and inspections throughout construction. While each community has its own approval process, all builders generally must follow the steps outlined below.

Before ground is broken, the builder submits plans for approval by the building department, the zoning or planning commission, and the health department if a septic system and well are planned. Final blueprints or working drawings are drawn up, and the builder files them with the building department to obtain a building permit [Figure 2]. Depending on the jurisdiction, a permit from the local environmental authority may also be required.

2. Sample building permit.

Project Address ______________________________

Assessor's Parcel No. ______________________________

Owner's Name ______________________ Phone __________

Address ______________________________

Arch/Eng. ______________________ License No. __________

Address ______________________________

Contractor's Name ______________________ Phone __________

Address (mailing) ______________________________

City License No. ______________________________

LICENSED CONTRACTOR DECLARATION

I hereby affirm that I am licensed under provisions of Charter 9 (commencing with section 7000) of Division 3 of the Business and Professions Code, and my license is in full force and effect.

License Class ______________________ License No. __________

Date ____________ Signature ______________________

WORKERS' COMPENSATION DECLARATION

I hereby affirm that I have a certificate of consent to self-insure or a certificate of Workers' Compensation Insurance, or a certified copy thereof (Section 3800, Lab. C.)

Policy No. ____________ Company ______________________

☐ Certified copy is hereby furnished.

☐ Certified copy is filed with the city Building Division.

Date ____________ Applicant ______________________

EXEMPTION FROM WORKERS' COMPENSATION DECLARATION

(This section need not be completed if the permit is for one hundred ($100) or less).

I certify that in the performance of the work for which this permit is issued, I shall not employ any person in any manner so as to become subject to the Workers' Compensation Laws of California.

Date ____________ Signature ______________________

NOTICE: If, after making this declaration, you should become subject to the Workers' Compensation provisions of the Labor Code, you must forthwith comply with such provisions or this permit shall be deemed revoked.

CONSTRUCTION LENDING AGENCY

I hereby affirm that there is a construction lending agency for the performance of the work for which this permit is issued (Section 3097 Civ. C.)

Lender's Name ______________________________

Lender's Address ______________________________

OWNER BUILDER DECLARATION

I hereby affirm that I am exempt from the Contractor's License Law for the following reason (Section 7031.5 Business and Professional Code: Any city or county which requires a permit to construct, alter, improve, demolish or repair any structure, prior to its issuance, also requires the applicant for such permit to file a signed statement that he/she is licensed pursuant to the provisions of the Contractor's License Law (Chapter 9 (commencing with Section 7000) of Division 3 of the Business and Professions Code) or that he/she is exempt therefrom and the basis for the alleged exemption. Any violation of Section 7031.5 by any applicant for a permit subjects the applicant to a civil penalty of not more than five hundred dollars ($500).

☐ I, as owner of the property or my employees with wages as their sole compensation, will do the work, and the structure is not intended or offered for sale (Section 7044, Business and Professional Code: The Contractor's License Law does not apply to an owner of a property who builds or improves thereon, and who does such work himself/herself or through his or her own employees, provided that such improvements are not intended or offered for sale. If, however, the building or improvement is sold within one year of completion, the owner will have the burden of proving he/she did not build or improve for the purpose of sale).

☐ I, as owner of the property, am exclusively contracting with licensed contractors to construct the project (Section 7044, Business and Professions Code: The Contractor's License Law does not apply to an owner of property who builds or improves thereon, and who contracts for such projects with a contractor(s) license pursuant to the Contractor's License Law). I am aware that proof of their Worker's Compensation insurance should be provided to me.

☐ I am exempt under Section ____________ B. & P. C. for this reason

____

Date ____________ Owner ______________________

I hereby certify that I have read this application and state that the above information is correct. I agree to comply with all city and county ordinances and state laws relating to building construction, and hereby authorize representatives of this city to enter upon the above-mentioned property for inspection purposes.

Signature ______________________ Date __________

Drivers' License or SS # ______________________

Applicant to fill in all spaces except those within heavy border lines and sign all appropriate declarations

PERMIT NO.	TRACT NO.	LOT NO.	UNIT NO.

TYPE OF CONSTRUCTION ☐ NEW ☐ ADD ☐ ALTERATION ☐ REPAIR
☐ CONVERSION ☐ DEMOLISH ☐ OTHER ______
DESCRIPTION OF WORK/USE OF BLDG. ______

TYPE OF PERMIT ☐ GRADING ☐ FOUNDATION ☐ STRUCTURAL
☐ TENANT ☐ ELECTRICAL ☐ PLUMBING ☐ MECHANICAL
☐ OTHER ______
PROJECT DESCRIPTION: Sq. Ft. ______ Occupancy ______
Construction type ______ Zone ______ Acres ______
Valuation $ ______ PC# ______

Remarks

A.P. #		Zone
SETBACKS	Main Building:	Accessory Building:
	Front	
	Rear	
	Right	
	Left	
Plng. Ref. #		Parking Req./Provided

ZONING REQMTS

Planning Notes

PLUMBING	FEES
Alter, Repair	
Auto Wash	
Backflow Preventer	
Backflow Irrigation	
Clarifiers	
Dishwasher	
Fixture Traps ea.	
Floor Drains	
Floor Sinks	
Gas Service Outlets	
Interceptors	
P-Trap	
Roof Drains	
Trailer Sewer	
Sewer Cap	
Sewer Connection - Bldg.	
Solar Collectors	
Solar Piping	
Solar Tank	
Swimming Pool/Spa	
Water Heater	
Water Service	
Septic System	
Misc.	
Issue	

ELECTRICAL	FEES
Generator ☐0-5KW ☐ 5-15KW	
Construction Pole Sub.	
Cooking Unit	
Pole Light	
Dryers ☐Gas ☐Electric	
Fan	
F.A.U.	
Fixtures	
Outlets	
Switches	
Meter ☐ up to 100 amps	
Meter ☐over 100 amps	
Sub panel	
Transformer ☐ 15-50KW ☐ 50+KW	
Motors ☐ 0-1 HP ☐1½ - 8	
Inc. Solar ☐9-15 HP ☐16 15+	
Sing/Fam. Res.	
Multi-Fam. Res.	
Swimming Pool/Spa	
Issue Fee	

MECHANICAL	FEES
Boiler	
Duct-Under Ground	
Duct-Structural	
Fan -Ventilating	
Exhaust System Multiple	
Evaporative Cooler	
Heating System & Ducting	
☐0-100,000 BTU ☐100,000 BTU	
Hood or Canopy	
Type 1 ☐ Type II ☐	
Refrigeration System	
☐0-100,000 BTU ☐100,000 BTU	
Firepl/Fact. Blt. ICBO #	
System Repair/Alteration	
Fire Dampers	
Registers	
Incinerator	
Misc.	
Issue	

SAMPLE

ACCT. #	GRADING FEES
	Cut: ______ cu. yards
	Fill: ______ cu. yards
3360-3370	Plan Check Fee $ ______
3300-3312	Permit Fee $ ______
	Total Fees $ ______
3360-3312	**Summary of Permit Fees**
	Building $ ______
	Plumbing $ ______
	Electrical $ ______
	Mechanical $ ______
	Other $ ______
110-231	SMIP $ ______
3360-3370	**Summary of Plan Check Fees**
	Building $ ______
	Plumbing $ ______
	Electrical $ ______
	Mechanical $ ______
	Other $ ______
3360-3375	Planning $ ______
	Total Fees Due $ ______

DEVELOPMENT SERVICES REQUIREMENTS:
Zoning Approved By ______ Date ______
Building Approved By ______ Date ______
Application Issued By ______ Date ______

Inspections

After permits have been issued and construction begins, inspections are required at specified stages of completion. The builder informs the appropriate department when the house is ready for inspection. The inspector then conducts the inspection and leaves a "passed" (or "failed") notice on the house. Only when the work has passed inspection can construction continue.

While inspection requirements vary from community to community, the following inspections are typical:

- Building or engineering department inspects municipal water service and sewer connections (unless well and septic system are to be installed).
- Building department inspects footings, open trenches and/or formwork before concrete is poured. If steel reinforcement is used, it is inspected at the same time. Footing depth and soil conditions are checked to ensure that the footings will provide adequate support for the structure above.
- Building department inspects foundation prior to waterproofing and backfilling.
- Health department inspects well and septic system.
- Building department inspects roughed-in framing, plumbing, electrical, heating and air conditioning systems, insulation and other items before walls are closed in.
- Building department performs final inspection to check plumbing, electrical, and mechanical systems, interior and exterior finish, and landscaping. If everything is in order, a certificate of occupancy is issued. (In some municipalities, the local board of fire underwriters must inspect the electrical installation before a certificate of occupancy can be issued.)

THE CONSTRUCTION PROCESS

To better understand how a house is built, let's examine the construction process from the ground up [Figure 3]. Note that the amount of time required to complete each phase of construction will vary depending on the design of the house, availability of materials, weather, and other factors.

3. Typical construction sequence.

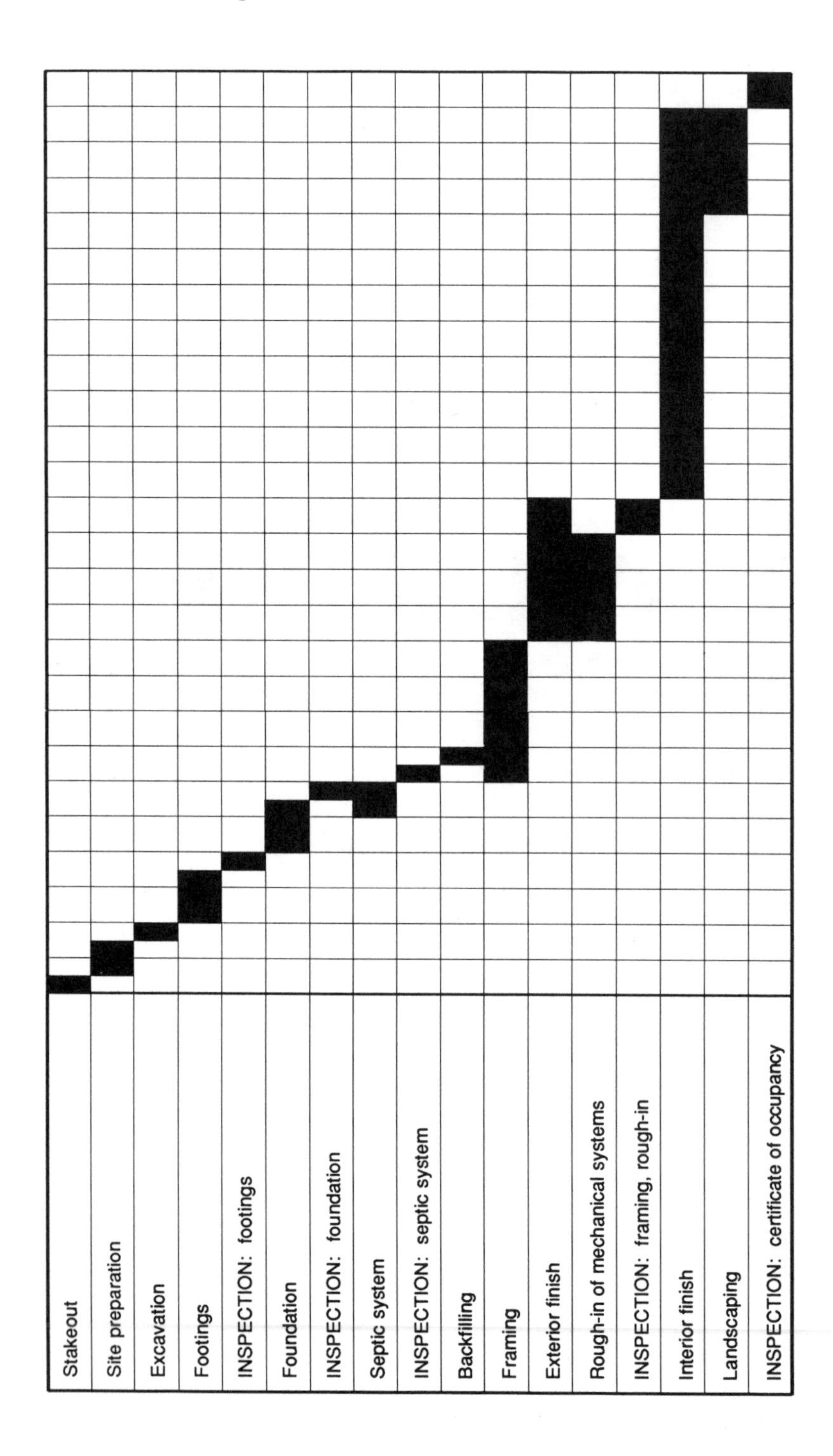

Stakeout

The process begins with a stakeout that positions the future house on the lot. Stakes are driven into the ground to mark the exact location of the house [Figure 4]. Placement of the house must conform to local zoning requirements. Existing site features such as terrain, trees, and rock outcroppings are also considered.

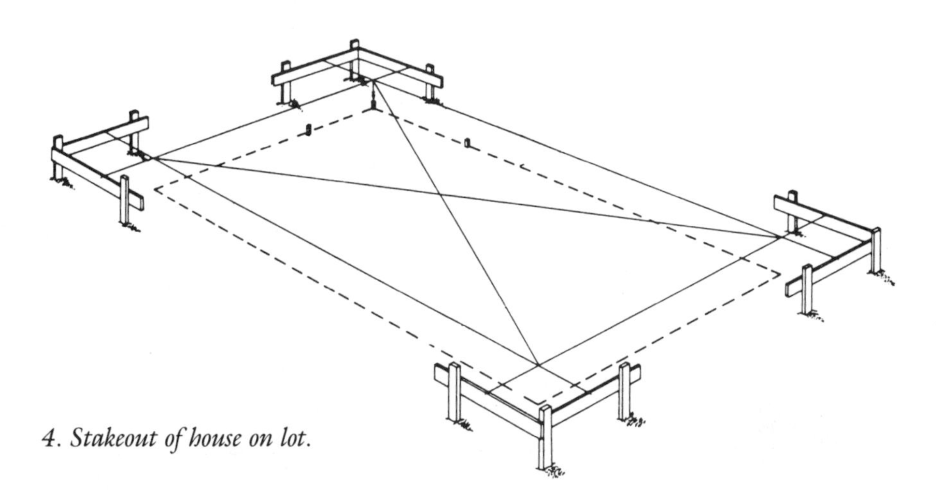

4. Stakeout of house on lot.

Site preparation

The next step is clearing and grading the site. This includes removing trees and changing the slope of the land where necessary, leveling, and providing access for the heavy trucks, equipment, and workers that will be serving the site throughout construction [Figure 5]. If a well is to be used, it is generally dug at this time.

Utility trench locations are marked for water, sewer and gas where available, and electrical service. (Note that electricity may be provided either underground or overhead via a utility pole connection.) The location of each utility's main, or connection point, dictates where trenches will be dug to the future house.

If a septic system is to be installed, a percolation test must be performed on the lot to determine the soil's water absorption rate and the best location for the system. Depending on the jurisdiction, the "perc" test is performed by a health department representative or an independent contractor such as a soils engineer.

5. Site clearing and grading.

Excavation

Once the site has been cleared and graded, a hole is dug (or "excavated") for the basement or crawl space if the house will have one [Figure 6]. The hole is made larger than the actual dimensions of the house to allow for working room and the installation of drain pipes along the base of the foundation. The bottom of the hole is leveled off in preparation for building the foundation walls that will support the house. Utility trenches are excavated at the depths and locations established during stakeout.

Many homes being built today do not have basements. Instead, they are built over crawl spaces or on concrete slabs. A crawl space is an unfinished area below the first floor of a house, usually just big enough to "crawl" through for maintenance and repair of ductwork and pipes. A crawl space requires excavation of a much shallower hole than does a full basement. Concrete floor slabs poured directly on prepared soil are called "slab on grade."

6. Excavation of basement.

Footings

Next come the footings. These form the base for the walls, chimney, and general structure of the house, whether or not the house has a basement. The footings extend beyond the walls to spread the weight of the house over a greater area, giving the structure more stability.

Poured concrete is the material most commonly used in constructing footings. A trench is dug around the edges of the basement hole where the foundation walls will be [Figure 7]. Concrete is then poured into the trench and leveled to form the footings [Figure 8]. Steel reinforcing bars are sometimes placed in the trench before the concrete is poured to give the footings added strength when soil conditions are poor. (If the house will have a treated wood foundation, gravel is used to spread the weight of the house instead of concrete footings.) Local codes usually govern the size of footings based on soil composition and anticipated building weight.

Depending on local building department requirements, a footing inspection may be required at this point.

7. Trench for footings.

8. Poured concrete footings.

Foundation

The foundation is built on top of the footings. It consists of walls that enclose the basement or crawl space and support the weight of the house above. Foundation walls can be built from concrete block [Figure 9], poured concrete that hardens between temporary wooden forms (called "cast-in-place") [Figure 10], or wood treated with a preservative to resist insects and rot.

Wall thicknesses and construction methods are determined by local building regulations or a structural engineer's calculations. The walls generally extend at least eight inches above ground level to protect the wood structure and finish materials from soil moisture and rot.

Exterior fireplace chimneys (those built on the outside of the house) are started when the foundation walls are built; the chimney goes up as the walls go up. If prefabricated chimneys are used, they are installed at the exterior finish stage.

The foundation walls are waterproofed with a black asphalt compound [Figure 11]. In poorly draining soils where heavier moisture protection is needed, "membrane" waterproofing (plastic sheeting or felt coated with asphalt or other bituminous materials) is applied to the foundation walls [Figure 12]. Clay or perforated plastic drain pipes are laid in gravel or crushed stone around the base of the foundation [Figure 13]. The pipes are sloped to carry water away from the foundation to a storm drain, to the ground surface, or to a sump pump located below the basement floor.

A building department inspection generally takes place upon completion of the foundation.

9. Concrete block foundation walls.

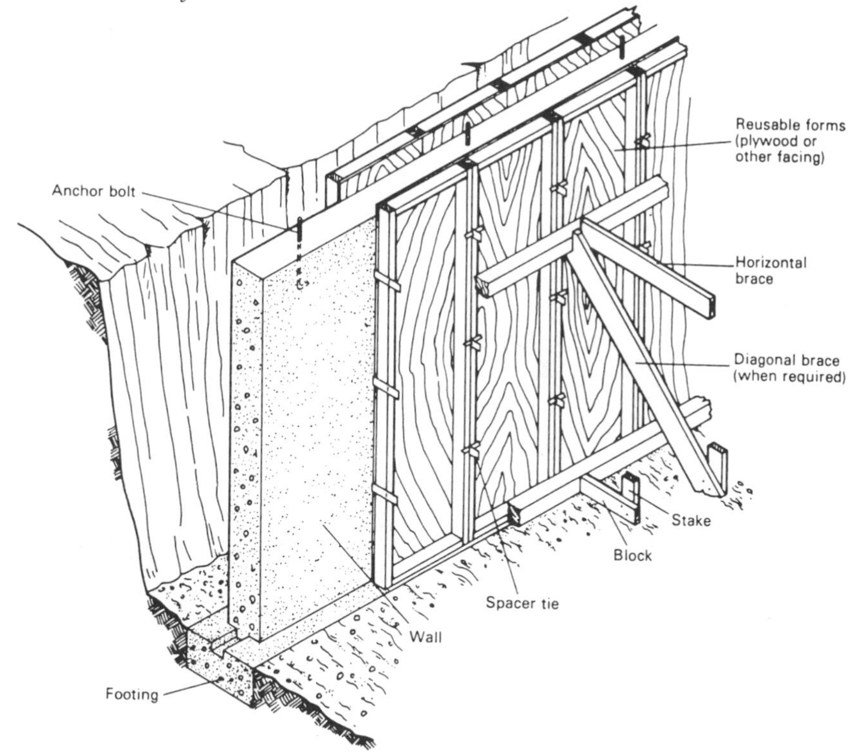

10. Poured concrete foundation walls.

11. Foundation walls with asphalt waterproofing.

12. Foundation walls with membrane waterproofing.

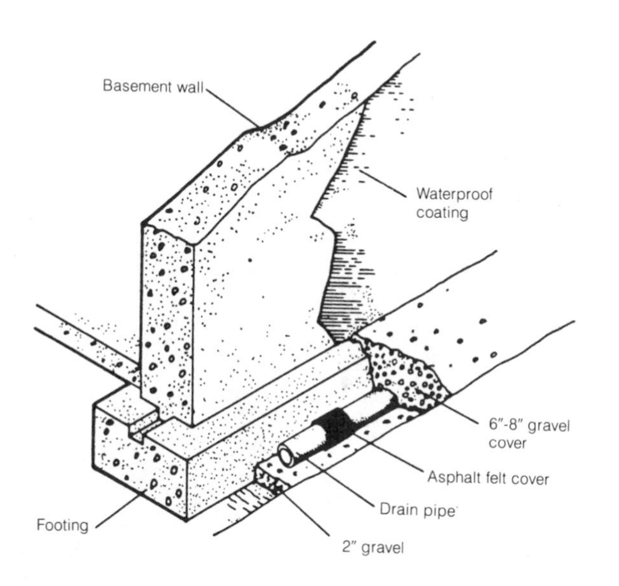

13. Drain pipe at base of outer foundation walls.

Septic system

A septic system is an onsite method of household waste disposal in which waste is processed for absorption into the soil. Septic systems are commonly used in less densely developed non-urban areas that are not served by public sewers.

Trenches for underground waste pipes are dug in the basement before the floor is laid [Figure 14]. Another trench is dug from the place where waste will leave the house to the designated septic (or "absorption") field. This is an area of the lot where the "perc" test has indicated that processed liquid waste may safely be absorbed into the soil. The septic field is bedded with gravel to aid the absorption process.

A septic tank is placed in the trench between the house and the septic field to collect the waste. Once solids have settled to the bottom of the septic tank, a pipe carries the liquid waste (or "effluent") to the septic field, where it gradually seeps through the gravel and into the soil [Figure 15].

A health department inspector will inspect the septic system to verify that it has been installed properly before the

14. Trenches for septic waste pipes.

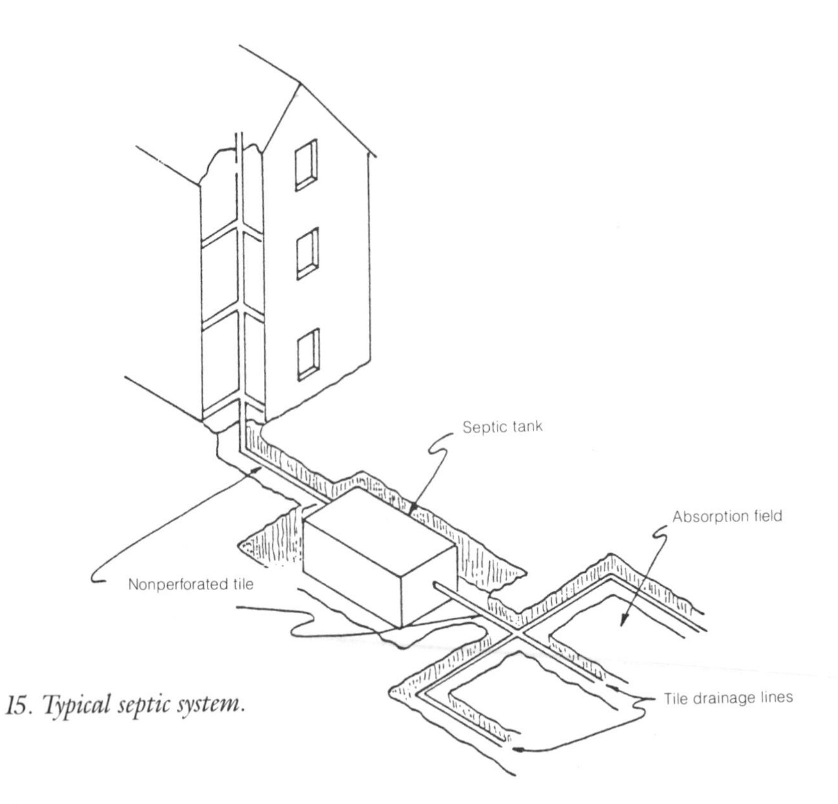

15. Typical septic system.

trench is filled in. Septic system maintenance is important. Septic tanks must be inspected and pumped out every two to four years. The frequency of servicing depends on household size, habits, garbage disposal use, and other use factors.

Backfilling

The earth is replaced in the trench around the foundation and the septic trenches; this is called backfilling. Termites can be a problem in most parts of the country, so the soil around a foundation should be treated with an approved pesticide.

Framing

Now floors, walls, ceilings, and the roof are framed with wood or steel studs, beams, and joists to provide a skeleton or internal structure for the house [Figure 16].

Floor framing—This is the first step. It provides a base for the walls and roof of the house. A horizontal wood or steel center beam is laid into the top of the foundation walls and supported between the walls by vertical wood or steel columns or piers [Figure 17]. Lengths of lumber called sills are fastened to the tops of the foundation walls. The sills are typically treated with a termite pesticide. Floor joists—parallel wood beams that support floors and ceilings—are nailed to the sills perpendicular to the center beam [Figure 18].

Sheets of plywood, flakeboard, or oriented strand board subflooring are nailed to the floor joists [Figure 19]. The subflooring becomes the base on which the finished floor (hardwood, carpet, vinyl, tile) is laid.

16. Framed house.

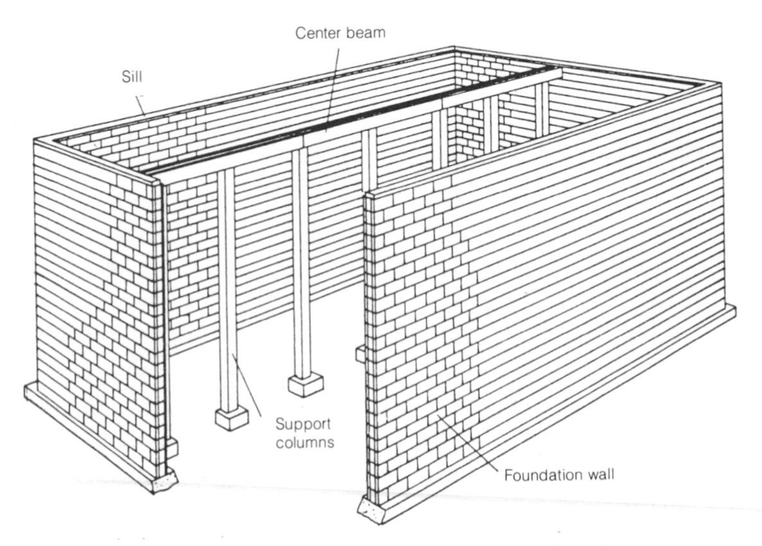

17. Center beam laid into foundation walls and supported by columns.

18. Floor joists.

19. Plywood subflooring.

Wall framing—Wall framing comes next. The most common framing method involves preassembly of wall sections on the subfloor. Sections are then tilted upright into position [Figure 20]. Wall sections consist of vertical wood or metal studs nailed into horizontal plates at top and bottom. Door and window openings are framed into the studs in preparation for installing the actual doors and windows [Figure 21]. Fireplace openings must also be allowed for during framing.

Exterior wall sheathing (panels of plywood, flakeboard, oriented strand board, or other material) is applied to reinforce the studs and provide a base for the exterior finish material. Some builders cover the sheathing with an air- and water-resistant polyethylene vapor barrier or treated building paper; others use rigid foam insulation board to keep out air and moisture [Figure 22].

20. Tilt-up method of wall framing.

21. Framed walls.

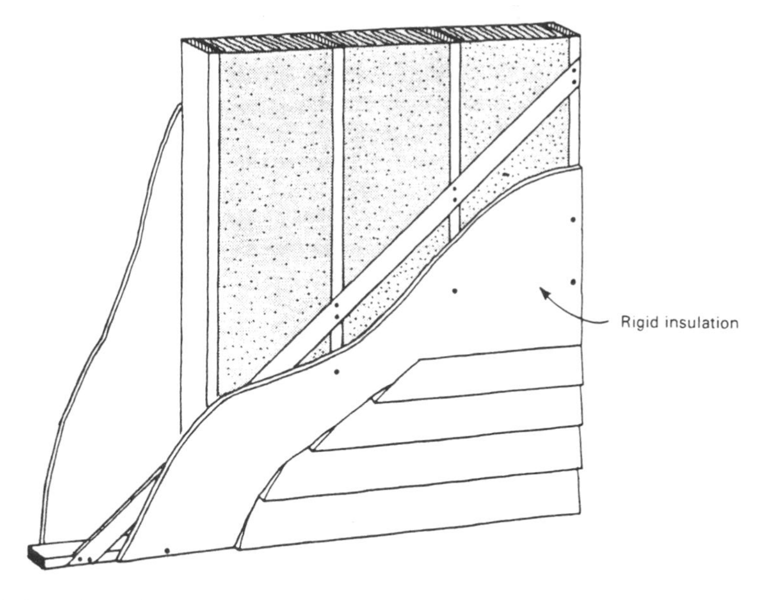

22. Wall sheathing with rigid foam insulation vapor barrier.

Ceiling and roof framing—Ceilings and the roof are framed once the walls are in place. Ceiling framing ties together opposite walls, supports the finished ceiling, and provides a base for a second story or attic storage area [Figure 23]. Ceilings are framed in much the same way as floors, using a horizontal series of joists.

Roof framing consists of three basic components: rafters, which support the weight of the roof; the ridgeboard, which forms the peak of the roof where the rafters join; and collar beams, which connect the rafters [Figure 24].

Many builders today use prefabricated framing members, called trusses, in place of conventional ceiling and roof framing. These triangular wood units are as strong as site-built ceiling joists and rafters, but are much easier to assemble. Trusses are delivered to the building site in quantities [Figure 25], and are lowered into position by crane [Figure 26].

Vents are installed in the roof to allow air to circulate freely through the attic. The vents help to remove condensation that could damage the internal structure of the house over time. Vents may be located under the eaves, in the gable end wall of the house, or along the roof ridge.

Next the roof is sheathed (commonly with plywood, flakeboard, oriented strand board, or one-inch lumber) to cover the rafters, provide structural strength, and serve as a base for the roofing material [Figure 27]. An underlayment of water-resistant roofing paper is usually applied to the sheathing to prevent moisture from seeping through. Flashing is installed at wall and roof intersections to channel water away from the structure [Figure 28].

23. Ceiling framing.

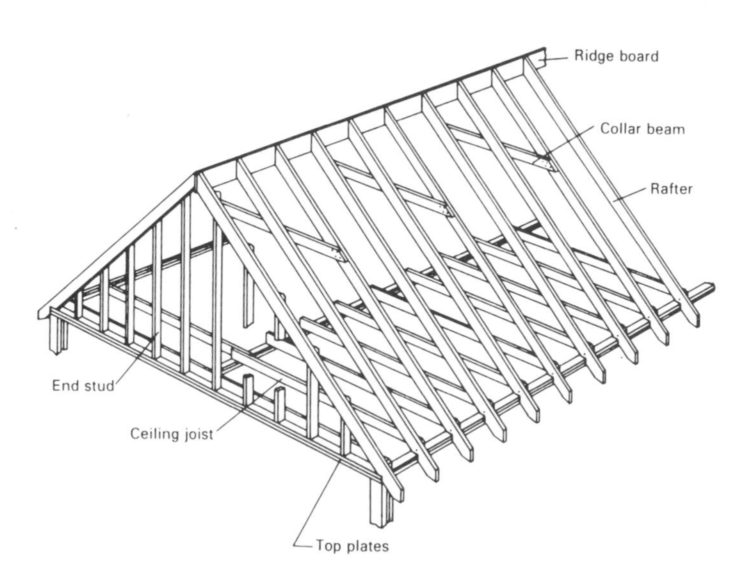

24. Roof framing.

25. Roof truss delivery.

26. Roof truss lowered into place by a crane.

27. Plywood roof sheathing.

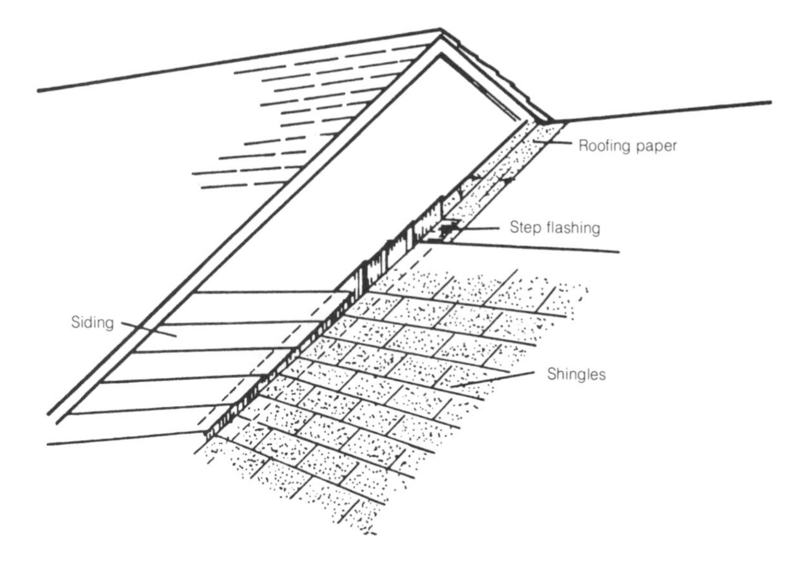

28. Roof underlayment, flashing, and shingles.

The roof is now ready for its finish layer. The choice of roofing material depends on budget, local codes, the design of the house, and individual preference. Options include wood or asphalt shingles and shakes, tile, slate, or sheet metals such as aluminum, copper, and tin [Figure 29].

Once the house is enclosed, shower and bathtub units are moved into place for hook-up later [Figure 30]. These oversize components will not fit through finished doorways. Depending on the type of unit being installed, the floor beneath the bathtub may be reinforced to support the extra weight. Similarly, special framing may be constructed to support the shower.

Windows and exterior doors go in next. They are prefabricated components that are delivered to the building site fully assembled for installation [Figure 31]. They can be constructed of wood, metal, vinyl, or metal or wood covered with vinyl.

29. Asphalt roof shingles.

30. Bathtub installation.

31. Door and window installation.

Exterior finish

The exterior wall sheathing is covered with a decorative—and protective—finish material such as siding (wood, vinyl, or aluminum) [Figure 32], brick or stone veneer, stucco, or a combination [Figure 33]. The choice of finish depends on budget, personal preference, and regional style.

Exterior trim is applied to finish the rough edges around windows and doors [Figure 34], cornices (also called eaves), gable ends, and porches. Gutters and downspouts are installed to channel runoff away from the house [Figure 35]. Trim material should be durable and weather-resistant. Wood, aluminum, and vinyl trim are commonly used. If a prefabricated chimney is called for, it goes up at this time.

32. Siding installation.

33. Combined exterior finish of aluminum siding and brick.

34. Exterior door trim.

35. Cornice trim and gutters.

Rough-in of mechanical systems

As finish is applied to the outside of the house, mechanical systems are installed or "roughed in" while interior wall cavities are open and accessible. These systems include plumbing lines, heating and air conditioning ducts, and electrical wiring, all of which will eventually be concealed behind the walls and under the floors. In areas where gas is available, plumbing for gas lines takes place during rough-in as well. Gas may be used to fuel the furnace, hot water heater, stove, and other appliances.

Plumbing—Plumbing fixtures (sinks and lavatories, showers, tubs, commodes, dishwashers, clothes washers, hot water heater) are often grouped to make more efficient use of pipes for water and waste handling. While the actual fixtures are installed as the house nears completion, the pipes and fittings that connect them are roughed in now.

Plumbing rough-in involves three separate networks of pipes: hot water, cold water, and waste. Hot and cold water pipes, which operate under pressure, usually run side by side until they reach faucets and appliances, where hot and cold water are mixed as they are used. Water pipes can be made of copper, galvanized steel, or a durable plastic called polyvinyl chloride ("PVC") [Figure 36].

Waste pipes operate by gravity rather than pressure, and must therefore be pitched to allow waste materials to flow through them. Waste systems require constant air circulation to prevent sewage decomposition within the pipes and the accumulation of insects and harmful sewage gases. Vertical ventilation pipes (or "stack vents") through the roof of the house introduce fresh air into the waste system and discharge gases to the outside. Waste systems also feature sealing devices called "traps" that prevent harmful gases and insects from entering the house through septic or public sewer connections. Below-ground waste pipes are made of asphalt-coated cast iron or PVC. Above-ground waste pipes may be cast iron, copper, or PVC, depending on local building code requirements.

36. Plumbing pipe installation.

Heating and air conditioning—Central heating and air conditioning systems are also installed at the rough-in stage. Heating systems generally use either the "forced air" or the "forced hot water" method of generating and circulating heat. Where central air conditioning is to be provided, a forced air system is more efficient because the same distribution network can be used for both heated and cooled air.

In a forced air system, air is heated by a centrally-located electrical, oil, or gas furnace [Figure 37]. A blower forces the heated air through a network of large metal or plastic passages, called ducts, to each room in the house [Figure 38]. Central air conditioning systems use the forced air method to carry cooled air through the ducts to outlets throughout the house.

Another forced air system is the heat pump, which is a single refrigeration unit that is used for both heating and cooling. It operates on electricity. Heat pumps follow the principle that outdoor air contains heat, or "thermal energy," even in winter. During the winter months, the heat pump draws in outdoor air, extracts the heat from it, and circulates that heat through the house. In the summer, the system is reversed. The heat pump removes heat from indoor air, discharges the heat outdoors, and circulates the cooled air through the house. Heat pumps tend to operate more efficiently in moderate climates.

In a forced hot water heating system, pipes carry heated water from a central gas- or oil-fueled boiler to radiators, baseboards, or other outlets throughout the house. Some forced hot water systems are designed to supply hot water for daily household use as well.

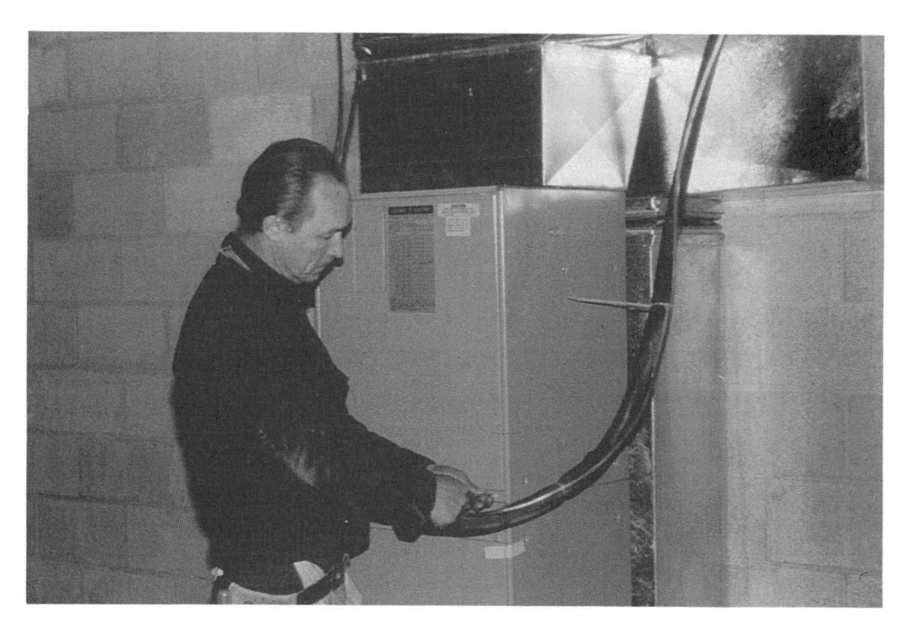

37. Forced air furnace installation.

38. Heating and air conditioning ducts.

Electrical—Electrical wiring is threaded through the studs and flooring that define each room [Figure 39]. All the wiring in the house passes through a metal box called a distribution panel, which is usually located at the utility company's hook-up point. The panel box distributes electricity via circuits, or sets of wires, to outlet receptacles in each room as well as the thermostat, light switches, doorbells, appliances, intercom, and security system [Figure 40].

The distribution panel contains a circuit breaker for each circuit in the house. Should the demand for electricity on a particular circuit become excessively high, the circuit breaker will automatically cut the flow of electricity through that circuit. The flow of electricity will remain broken until the homeowner resets the circuit breaker. This safety feature, which is required by code throughout the country, helps to reduce electrical hazards in the home. The ground fault interruptor is another type of circuit breaker that is used to reduce the danger of electrical shock in areas where water is commonly found (such as the kitchen, bath, and outdoor areas).

39. Electrical wiring.

40. Distribution panel.

Insulation

Once the mechanical systems have been roughed in, insulation is placed in floors, ceilings, foundation walls, aboveground walls, and in the attic—all the places through which air is likely to seep. Insulation comes in several forms: flexible mineral wool batts and blankets (used in walls, floors, ceilings, and around ducts), semi-rigid and rigid plastic or fiberglass panels (used primarily in walls), and loose fiberglass, mineral wool, or cellulose fill that can be blown or poured into place (used in walls and ceilings) [Figure 41]. Plastic foam is also used for spot insulation around windows and doors, pipe openings, and other air leakage points.

The term "R-value" is commonly used in reference to insulation. When used with a number (such as R-19), it indicates the level of resistance to heat flow in a building material. The higher a material's R-value number, the more effective insulation it provides.

The house is now a maze of wires winding through studs, shiny ducts for heating and cooling, plumbing pipes and

41. Types of insulation.

Batt

Rigid panel

Loose fill

Courtesy Owens-Corning Fiberglas Corporation

insulation of all kinds. This is the time for the next building inspection. Because roughed-in systems and insulation will be hidden behind walls and under flooring, the inspector must determine that each is installed according to code and that the framing has been built properly.

Interior finish

Now that all roughed in systems have passed inspection, the interior walls and floors may be finished off. Trim is applied, appliances are installed, and the house nears completion.

Basement floor—With all the necessary underground utility connections in place, the basement floor is installed. A poured concrete slab is generally used. A layer of coarse gravel is laid before the slab is poured to keep excess ground moisture from seeping into the basement. A vapor barrier of six-mil polyethylene film is often used on top of the gravel base for additional moisture protection. One or more floor drains (depending on basement size) should be installed in the basement floor prior to pouring the concrete.

The concrete is poured so that the slab slopes slightly toward the floor drain. The wet concrete is screeded (leveled off) with a length of lumber, then given a smooth finish [Figure 42].

The ground floors of houses built slab on grade (without basements) are installed in much the same way as basement floors.

Drywall—The insulation is covered with sheets of drywall (also called gypsum wallboard), which form the interior walls and ceilings of most houses. The sections of drywall are taped at the seams, coated (or "spackled") with a joint compound and allowed to dry [Figure 43], then sanded to prepare them for painting, wallpapering, or the application of textured spray. This job requires meticulous attention to prevent waves and joints from appearing in the finished walls. Drywall is also used as a base for paneling and masonry finishes.

Other finish work—Once the dust from insulation and drywall installation has settled, the remaining interior finish work takes place. Interior doors are hung [Figure 44]. Mold-

42. Poured concrete basement floor: leveling and finishing.

43. Taped and spackled drywall.

44. Hanging interior doors.

ing is applied around windows and doors [Figure 45], and can also be used where walls, floors and ceilings meet. Chair rails may be applied to walls as a decorative and protective trim. Fireplaces are finished with a face of brick, stone, marble or tile, with or without a mantel. Rooms are painted or wallpapered, and paneling goes up. Wood and vinyl finish floors go in (other types of finish flooring, such as tile and wall-to-wall carpeting, are usually laid later, after cabinets and fixtures have been installed).

Where wood flooring will be used, an underlayment of treated building paper is commonly laid over the subflooring. The underlayment helps to block dust and moisture from below and reduces squeaks by separating the finish floor from the subfloor. Both prefinished and unfinished wood floors are available from suppliers. If prefinished flooring is used [Figure 46], it must be installed with special care and its surface should be protected with a drop cloth until all remaining interior finish work is complete. If unfinished wood floors are installed, they are sanded and varnished after all other interior finish work is done.

45. Interior door trim molding.

46. Installing prefinished wood flooring.

47. Installing vinyl flooring.

Vinyl flooring is commonly used in kitchens and bathrooms [Figure 47]. It is installed over an adhesive. Because its surface is resilient and therefore susceptible to marking, extra care should be taken when moving furniture and equipment over it. Once the vinyl flooring is in place, kitchen cabinets and appliances go in [Figure 48]. Bathroom vanities, sinks, and commodes are installed and medicine cabinets are hung. Then tile floors can be laid.

Floor tile can be of several types—ceramic, clay, mosaic, or marble. The method of installation depends on the type of tile being used. Tile is also used on wall surfaces, particularly in bathrooms and kitchens [Figure 49].

Base molding is nailed in. Also called baseboards, this type of trim serves as a decorative and protective finish between the walls and the floor.

Now hardwood floors receive their final sanding and varnishing. This can take a week or more, as the floor must dry completely between each application of varnish.

48. Installing kitchen cabinets and appliances.

49. Ceramic wall tile application.

Wall-to-wall carpeting goes in after all other interior finish work is completed. It is installed over padding, which is nailed, stapled, or glued to the subfloor. Sections of carpet are pulled tight to the corners of the room as they are laid in place [Figure 50], and the pieces are joined with adhesive tape. An iron is used to melt the adhesive so that the seams cannot be seen.

Finish plumbing and electrical—While the plumbing and electrical systems were roughed in before the insulation and drywall went up, additional finish work is needed to prepare the home for occupancy. Appliances—washers and dryers, dishwashers, refrigerators—are hooked up, as are plumbing fixtures and the hot water heater [Figure 51]. Light fixtures, switch and outlet receptacle covers are installed [Figure 52]. All electric, water, and gas connections are tested to ensure that they are leak-free. Then hook-ups to outside utility services are made.

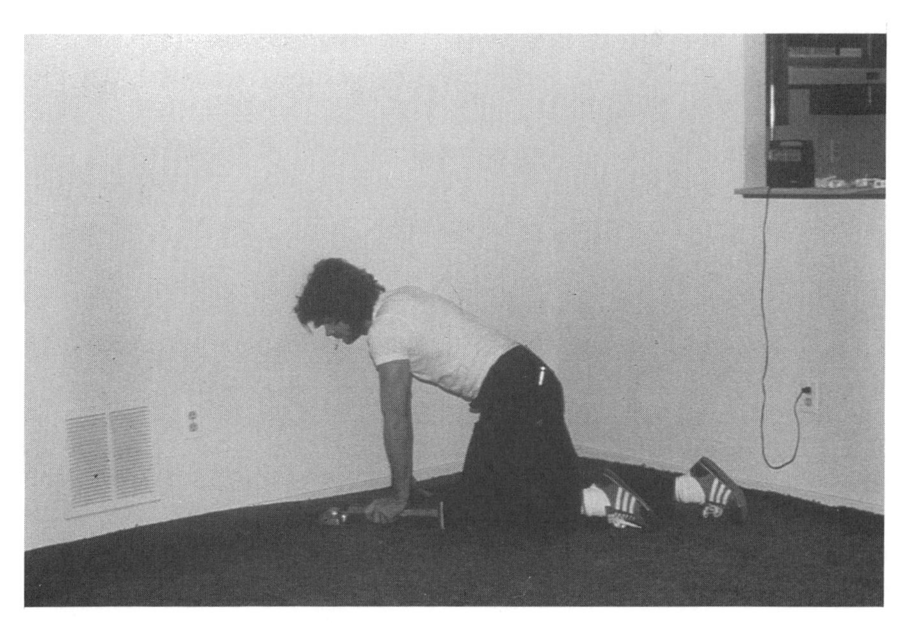

50. Installing wall-to-wall carpeting.

51. Finish electrical: hot water heater hook-up.

52. Finish electrical: light fixture installation.

LANDSCAPING

House construction is now complete. The delivery trucks have come and gone and the trash dumpsters have been hauled away. Now the exterior of the house receives its finishing touches in the form of landscaping. This is a general term for final site grading and the installation of decks and patios, walkways and driveways, lawns and gardens, shrubs and trees.

Decks and patios are design options that extend a home's outdoor living space. Their size and shape are determined by the configuration of the house and yard, budget, personal preference, and any applicable local zoning ordinances. Decks are elevated outdoor platforms, usually constructed of wood and edged by a railing for safety and convenience [Figure 53]. Patios are built directly on the ground, often over a gravel base. Common patio surfaces include brick, flagstone, and poured concrete.

Site grading prepares the lot for walkways and driveways, lawns and plantings. The lot is smoothed using equipment such as a bulldozer and a backhoe, and the earth left from excavating the basement is redistributed on the lot [Figure 54]. The earth is sloped away from the house for more effective drainage. Sloping must be maintained by the homeowner to minimize future seepage problems.

53. Deck construction.

54. Final site grading.

Once the lot has been graded, driveways and walkways go in. Driveways can be long or short, straight, curved or circular depending on the size and shape of the house and lot. They can be constructed of asphalt, gravel, crushed stone, or concrete [Figure 55]. Walkways are installed to link the house with the street and driveway, and to provide a path around the house. Concrete, brick, flagstone, and asphalt are commonly used.

After the paving is finished, lawns are started using grass seed or pregrown mats of grass called sod. Shrubs and trees may be planted at this time for decoration, privacy, shade, and wind protection [Figure 56].

55. *Asphalt driveway construction.*

56. *Planting shrubs.*

CERTIFICATE OF OCCUPANCY

Before the house can be occupied, the local building department must perform a final inspection of plumbing, heating and air conditioning, and electrical systems, interior and exterior finish work. Once the building inspector determines that the house has been built in compliance with all applicable code and zoning requirements, a certificate of occupancy is issued [Figure 57].

CITY OF OAKMONT

CERTIFICATE OF

COMPLETION AND OCCUPANCY

Building Permit No.____________________ Date Issued__________

Property Address____

Lot No._____ Block No._____ No. Handicapped Units_____

Zoning District_____ Conforming_____ Legal Non-conforming_____

Owner, Occupant, or Contractor __________________________________

Use of Property____

Comments____

____

BUILDING CERTIFIED IN COMPLIANCE AS OF__________________

This is to certify that all required inspections have been made and that the work authorized under the above-numbered Building Permit has been completed in compliance with the building law.

If the Certificate of Completion and Occupancy is issued by direction of the Zoning Board of Adjustment, it shall be subject to any and all conditions specified or any part thereof shall be held void or invalid, or if any such conditions are not complied with, this Certificate shall be void and of no effect.

This Certificate of Completion and Occupancy shall remain in force until such time as there is an alteration, addition or change of use in the structure, or land, when it shall become void and a new certificate must be issued.

DATE ISSUED____________________

BUILDING INSPECTOR____________________

57. Sample certificate of occupancy.

GLOSSARY

Asphalt A dark, tarlike material commonly used in the building industry for such uses as roofing, waterproofing and dampproofing, exterior wall covering, and pavement. A bituminous material.

Backfill Earth or other material used to fill in around foundation walls, usually built up to drain water away from the foundation.

Base molding A decorative band of finish board used to cover the joint between wall and floor. Also called baseboard.

Bituminous Containing tar, asphalt, or pitch. Bituminous materials are effective in retarding air and moisture flow.

Blueprints Complete construction plans, drawn to scale, used by builders and subcontractors to build a structure. They usually include site plan, foundation plan and cross-section, floor plans, elevations, building and wall cross-sections, mechanical systems, and special construction details. The term blueprint derives from the chemically treated blue paper on which drawings are printed in white; the term commonly refers to any set of working drawings, whether printed on blue, white, or other paper.

Building code Minimum legal requirements for all aspects of construction, established and enforced by local governments to protect public health and safety. Codes are established by building officials and others with first-hand knowledge of construction practices.

Building paper Heavy paper treated with bituminous material to retard air and moisture flow. Used as an underlayment between exterior sheathing and finish materials, and between subflooring and finish flooring.

Cast-in-place A term used to describe concrete that is poured between wooden forms to harden in the place where it is needed.

Cellulose Recycled wood fiber used in the building industry for insulation and flooring material. Can be chemically treated with flame retardant.

Center beam A wood or steel member that runs the length of the first floor of a house, bearing on the foundation wall at each end of the house and supported along its length by columns or piers. The center beam supports the house structure above it.

Certificate of occupancy A legal document issued by a building inspector, stating that a house has passed all inspections and is ready for utility hook-up and occupancy.

Chair rail A band of molding applied at chair back height along a wall to protect the wall finish from chairs being pushed against it; also used as a decorative detail.

Circuit breaker A safety feature for each electrical circuit in the distribution panel. Should the demand for electricity on a particular circuit be excessive, the circuit breaker automatically cuts the flow of electricity through that circuit. The electrical circuit remains broken until the circuit breaker is reset.

Collar beam In roof framing, a horizontal piece that provides structural strength by connecting opposite rafters.

Commode Water closet; toilet.

Concrete Mixture of cement, sand, gravel, and water that hardens into a rocklike mass.

Concrete block Precast hollow or solid building block made of cement, water, and aggregate such as sand, gravel, or crushed stone. Commonly used in wall construction.

Cornice On the exterior of a house, structural trim where the roof and walls meet. Also called eaves when the trim overhangs the walls.

Crawl space In houses without basements, the space between the ground surface and the first floor, made big enough to "crawl around in" for utility installation and repairs.

Crown molding A decorative band of finish board used to cover the joint between wall and ceiling.

Cultured marble A manufactured marblelike material commonly used for countertops and lavatory surfaces. It is a cast polyester resin mixed with crushed marble, then molded, cut, and polished. It is water resistant, and is lighter weight and less expensive than quarry marble.

Distribution panel Metal box through which all electrical wiring passes, usually located at juncture with utility company line. Electricity is distributed from the panel to all usage points by means of circuits, or sets of wires.

Dormer A projection built out from a sloping roof as a room extension or for a window.

Drain pipe Clay or plastic pipe, sometimes perforated, laid in the ground to carry fluid away from saturated areas and disperse it. Also called drain tile.

Drywall General term for a type of interior wall construction using "dry" gypsum wallboard panels instead of plaster. Also called wallboard.

Duct, ductwork Round or rectangular sheet metal or vinyl passages used to transfer heated and cooled air from heat and air conditioning sources to the various rooms in a building.

Eaves That part of a roof that extends beyond the walls of a building.

Effluent Liquid waste discharged from a septic or sewage treatment system.

Excavation Removal of earth or rock to create a hole, as for the basement of a house.

Fiberglass A nonflammable material made of spun glass fibers. It is used in thick woollike blankets as building insulation; woven into fabrics; and used to reinforce plastic resins in durable, molded, solid forms for a variety of uses.

Fire underwriters In some communities, a local board that performs inspections of electrical work to ensure compliance with the electrical code.

Flagstone Flat pieces of stone used as a paving surface for patios, walkways, and steps.

Flakeboard A plywood substitute manufactured from wood flakes and a resin binder pressed into boards.

Flashing Sheet metal or plastic used to cover joints and openings in exterior surfaces to protect against water leakage.

Footing Widened support, usually concrete, at the base of foundation walls, columns, piers, and chimneys. Designed to distribute the weight of these elements over a larger area and prevent uneven settling.

Formwork Support structure for freshly poured ("cast-in-place") concrete.

Foundation Walls, partially below-ground, that support the weight of the building above and enclose the basement or crawl space.

Framing The process of constructing the internal skeleton of a structure, usually of wood or steel studs, beams, and joists; also, the term used to describe the internal skeleton.

Gable The triangular end wall of a building that extends from the eaves to the peak of the roof.

Grading The preparation of a site by digging, filling in, or both, to accommodate construction of a building. Also, filling in with earth or other material around a completed building, at a slope to direct rain water runoff away from the building.

Gutter Metal, plastic, or wood channel at the eaves of a building, sloped slightly to carry off rain water and snow melt.

Gypsum board Panels used in drywall construction, consisting of the mineral gypsum pressed between two layers of heavy paper. Also called drywall, wallboard.

Heat pump A forced air heating and air conditioning system. In winter, heat is extracted from air and circulated through the house. In summer, heat is extracted and discharged outdoors, and the cooled air is circulated through the house.

HVAC Common building industry abbreviation for heating, ventilation, and air conditioning systems.

Inspection Examination of work completed on a structure to determine compliance with building code and other code requirements.

Insulation Any material used in building construction to resist heat loss, protect against sound transmission or fire, or to cover electrical conductors.

Joint compound A pastelike material used to cover tape at drywall seams for a smoother finish. Also called spackle.

Joists A series of horizontal parallel beams that support floors and ceilings.

Lavatory Wash basin; commonly called a sink.

Load bearing Providing support for a building's weight.

Masonry General construction term for materials set in mortar, including stone, brick, concrete, tile, and glass block.

Mechanical systems General term for plumbing, heating and air conditioning, and electrical systems.

Membrane waterproofing Method of protecting a structure against moisture using plastic sheeting or felt layers coated with asphalt or other bituminous material.

Mineral wool A loose, fibrous insulation material made from rock and molten slag.

Molding Wood, metal, or plaster strips used for decorative finish around windows and doors, at the top and base of walls, and along cornices.

Mortar A thick, pastelike material that hardens to bond masonry units together. Usually made of a mixture of cement, lime, sand, and water.

National Electrical Code (NEC)® Nationally accepted requirements used to regulate electrical installations. Established and administered by the National Fire Protection Association.

Oriented strand board A plywood substitute composed of layers bonded together with resin. Each layer consists of compressed strands of wood fiber oriented in a single direction; layers alternate direction of strand orientation.

Percolation test A soil test used to determine the rate at which water will be absorbed into the ground. Results are used to establish best locations for septic fields on a piece of property and to determine their size. Also called "perc test."

Permit A document issued by a local government agency allowing construction work to be performed in conformance with local codes. Work may not commence until permits have been obtained, and each permit-issuing agency must inspect the work at certain specified points during construction.

Plywood A type of building material made by gluing three or more thin layers (or "plies") of wood together in panels. Plies are laid so that the wood grain alternates direction with each layer; this increases the plywood's overall strength and counteracts warping in each ply.

Polyethylene A durable, pliable, waterproof plastic film used in construction as a vapor barrier.

Polyvinyl chloride (PVC) Rigid, durable plastic material used in plumbing for pipes and fittings.

Quarry marble Marble that has been extracted from a naturally-occurring land source and has been cut and polished for use in construction.

R-value A term which, when used with a number, indicates the level of resistance to heat flow in a building material. The higher a material's R-value, the more effective insulation it provides.

Rafter One of a series of structural members that form the legs of the triangle created in roof framing; joined at the peak of the triangle by the ridgeboard. Rafters support roof sheathing and finish materials.

Ridgeboard The length of lumber at the peak of a roof; supports upper ends of the rafters.

Roofing paper See building paper.

Rough-in The stage of construction that follows framing, when installation of all systems that will be concealed behind the walls—plumbing, heating and air conditioning, and electrical wiring—occurs.

Screed To level off freshly poured concrete and plaster. Also, the straight-edged device used in the screeding process.

Septic system A sewage disposal system for individual homes. A holding tank for raw sewage is installed in the ground, where sewage is broken down and liquefied by bacterial action. Small amounts of solid matter do not break down, but settle to the bottom of the tank. (The tank must be cleaned out every few years.) Liquid waste is discharged to a distribution or absorption field where it slowly passes into the soil and is purified.

Setback The minimum allowable distance between a structure and its lot lines, established by local zoning ordinances.

Sewer A system of pipes for carrying away storm runoff, waste water, or sewage to a municipal processing plant.

Shake Hand-split wood shingle.

Sheathing Sheets of plywood, flakeboard, oriented strand board, or insulation board used to cover the exterior of a building's frame.

Shingles Roof or wall covering of asphalt, wood, tile, slate, or other material cut into standard lengths, widths, and thicknesses.

Siding The exterior finish of a house applied over the sheathing; generally wood, plastic coated wood, vinyl, or aluminum.

Sill A support member laid flat on the top of the foundation wall, used as the base for floor framing; also called the sill plate. Also, the member forming the lower side of an opening, such as a windowsill or doorsill.

Slab A flat layer of poured concrete.

Sod Top layer of soil, containing grass and grass roots. Available in precut mats for starting lawns.

Soffit Exposed underside of a projecting building part such as a cornice or eaves.

Spackle A patching compound used to fill plaster or drywall cracks or nail holes; also called joint compound.

Stack vent Vertical waste ventilation pipe in a plumbing waste system. Discharges gases and introduces fresh air into the system.

Stakeout Measuring of house dimensions on a lot in accordance with the house plans and using stakes to indicate each corner.

Stucco A plaster cement used as an exterior and interior wall finish.

Stud Upright wood or metal members used as supporting elements in walls and partitions.

Subcontractor A person or company that contracts with the builder to perform work on a specific part of a construction job, such as excavation, plumbing, electrical work, or landscaping. Also called "sub."

Subflooring Rough boards, plywood, flakeboard, or oriented strand board laid on top of the floor joists, to which the finish floor is fastened.

Sump pump Device used to remove liquid from a drainage pit (or "sump").

Termite A wood-devouring insect that can demolish the woodwork of a structure.

Transformer A device for increasing or decreasing electrical current. Used by electrical utilities to convert high voltage levels for use in individual buildings.

Trap In plumbing, a bend in a waste pipe designed to hold water. The water acts as a seal to prevent insects and harmful sewage gases from discharging into plumbing fixtures inside the house.

Trench A narrow excavation in the earth for the installation of footings, pipes, drains, and electrical cables.

Trim Interior finish materials, including door locks, knobs, hinges, and other metal hardware; moldings around windows and doors; and other decorative work.

Truss Preassembled roof framing member, fabricated of wood, commonly manufactured in a triangular configuration that replaces ceiling joists, rafters, and collar beams.

Underlayment Moisture resistant material, such as an asphalt-treated paper, applied over roof and wall sheathing and under roof and exterior finish to prevent water from entering the structure. Also used between subflooring and finish floors.

Utilities Public services available to all citizens of a community, such as water, electricity, gas, and sewage disposal.

Vanity A dressing table; also, a wash basin with an enclosed cabinet below.

Vapor barrier Treated paper or plastic film that retards the flow of air and moisture.

Veneer Any decorative, nonstructural surface layer.

Wallboard See gypsum board, drywall.

Water closet Commode; toilet.

Zoning Division of a county or municipality into land use categories. Establishment of regulations governing the use, placement, spacing, and size of land parcels and buildings in each category.

APPENDIX

State builders associations affiliated with the National Association of Home Builders

The National Association of Home Builders (NAHB) is a federation of more than 800 state and local builder associations representing over 155,000 builders and related industry professionals nationwide. State and local builders associations are an excellent source of information about builders, house construction, and home buying practices.

The following is a list of state builders associations, listed in alphabetical order by state. Your state association can refer you to the local builders association in your area. You can also contact the National Association of Home Builders for additional information about local builders associations. (National Association of Home Builders, 15th and M Streets, N.W., Washington, D.C. 20005, 202/822-0200.)

Home Builders Association of Alabama
Montgomery, AL 205/272-1024

Alaska Home Builders Association
Anchorage, AK 907/248-7776

Home Builders Association of Central Arizona*
Phoenix, AZ 602/274-6545

Flagstaff League for Advancing Good Growth*
Flagstaff, AZ 602/774-4589

Southern Arizona Home Builders Association*
Tucson, AZ 602/795-5114

Arkansas Home Builders Association
Little Rock, AR 501/663-1428

*Arizona has no state builders association.

California Building Industry Association
Sacramento, CA 916/443-7933

Colorado Association of Home Builders
Denver, CO 303/753-0601

Home Builders Association of Connecticut
Hartford, CT 203/247-4416

Home Builders Association of Delaware
Wilmington, DE 302/994-2597

District of Columbia Building Industry Association
Washington, D.C. 202/966-8665

Florida Home Builders Association
Tallahassee, FL 904/224-4316

Building Industry Association of Georgia
Atlanta, GA 404/763-2453

Building Industry Association of Hawaii
Honolulu, HI 808/847-4666

Idaho Building Contractors Association
Boise, ID 208/377-3552

Home Builders Association of Illinois
Springfield, IL 217/753-3963

Indiana Builders Association
Indianapolis, IN 317/236-6334

Home Builders Association of Iowa
Des Moines, IA 515/278-0255

Home Builders Association of Kansas
Topeka, KS 913/233-9853

Home Builders Association of Kentucky
Frankfort, KY 502/875-5478

Home Builders of Louisiana
Baton Rouge, LA 504/387-2714

Home Builders Association of Maine
Augusta, ME 207/622-4990

State of Maryland Institute of Home Builders
Annapolis, MD 301/261-2997

Massachusetts Home Builders Association
Boston, MA 617/720-2340

Michigan Association of Home Builders
Lansing, MI 517/484-5933

Builders Association of Minnesota
Saint Paul, MN 612/646-7959

Home Builders Association of Mississippi
Jackson, MS 601/969-3446

Home Builders Association of Greater Kansas City**
Kansas City, MO 816/942-8800

Home Builders Association of Greater St. Louis**
St. Louis, MO 314/994-7700

Montana Home Builders Association
Billings, MT 406/259-1346

Nebraska State Home Builders Association
Lincoln, NE 402/474-6947

Nevada Home Builders Association
Las Vegas, NV 702/870-7234

Home Builders Association of New Hampshire
Concord, NH 603/228-0351

New Jersey Builders Association
Plainsboro, NJ 609/275-8888

New Mexico Home Builders Association
Albuquerque, NM 505/344-7072

New York State Builders Association
Albany, NY 518/465-2492

North Carolina Home Builders Association
Raleigh, NC 919/833-4613

North Dakota Association of Builders
Bismarck, ND 701/222-2401

Ohio Home Builders Association
Columbus, OH 614/228-6647

Oklahoma State Home Builders Association
Oklahoma City, OK 405/843-5579

Oregon State Home Builders Association
Salem, OR 503/378-9066

Pennsylvania Builders Association
Harrisburg, PA 717/234-6209

**Missouri has no state builders association. Either of these associations can refer you to the local builders association in your area.

Home Builders Association of Puerto Rico
Santurce, PR 809/723-0279

Rhode Island Builders Association
Providence, RI 401/521-0347

Home Builders Association of South Carolina
Columbia, SC 803/771-7408

Home Builders Association of South Dakota
Sioux Falls, SD 605/361-8322

Home Builders Association of Tennessee
Nashville, TN 615/726-1700

Texas Association of Builders
Austin, TX 512/476-6346

Home Builders Association of Utah State
Salt Lake City, UT 801/268-8750

Home Builders Association of Vermont
Rutland, VT 802/773-6251

Home Builders Association of Virginia
Richmond, VA 804/643-2797

Building Industry Association of Washington
Olympia, WA 206/352-7800

Home Builders Association of West Virginia
Charleston, WV 304/342-5176

Wisconsin Builders Association
Madison, WI 608/249-9912

Wyoming Home Builders Association
Cheyenne, WY 307/635-6467

ADDITIONAL READING

Alternatives to Public Sewer. Washington, D.C.: National Association of Home Builders, 1978.*

Basic Plumbing, Sunset Books. Menlo Park, CA: Lane Publishing Co., 1983.

Building Your Own Home, by Wasfi Youssef. New York: John Wiley & Sons, 1988.

Dreams to Beams: A Guide to Building the Home You've Always Wanted, by Jane Moss Snow. Washington, D.C.: National Association of Home Builders, 1989.*

Dwelling House Construction, by Albert G.H. Dietz. Cambridge, MA: MIT Press, 1974.

Estimating for Home Builders, by John C. Mouton. Washington, D.C.: National Association of Home Builders, 1988.*

How to Design and Build Your Own House, by Lupe DiDonno and Phyllis Sperling. New York: Alfred A. Knopf, Inc., 1984.

Land Development. Washington, D.C.: National Association of Home Builders, 1987.*

Residential Construction Drawings, by Mark W. Huth. New York: Van Nostrand Reinhold Company, Inc., 1983.

*For more information about books published by the National Association of Home Builders, call the Home Builder Press at 800/368-5242 or write the Home Builder Press, National Association of Home Builders, 15th and M Streets, N.W., Washington, D.C. 20005.

Residential Wastewater Systems. Washington, D.C.: National Association of Home Builders, 1980.*

Understanding Building Codes and Standards in the United States. Washington, D.C.: National Association of Home Builders, 1986.*

Wood Frame House Construction. Washington, D.C.: National Association of Home Builders, 1988.*

*For more information about books published by the National Association of Home Builders, call the Home Builder Press at 800/368-5242 or write the Home Builder Press, National Association of Home Builders, 15th and M Streets, N.W., Washington, D.C. 20005.

INDEX

NOTES